WHY IS VIRGIN ORBIT LAUNCHING ROCKETS FROM A PLANE
A Journey Beyond Tradition

Unlocking a New Era of Fast, Flexible, and Affordable Space Access

Tommy S. Manley

Copyright © **Tommy S. Manley, *2024*.**

All rights reserved. No part of this publication may be reproduced, distributed, or transmitted in any form or by any means, including photocopying, recording, or other electronic or mechanical methods, without the prior written permission of the publisher, except in the case of brief quotations embodied in critical reviews and certain other noncommercial uses permitted by copyright law.

Table of Contents

Introduction...4
Chapter 1: The Rise of Small Satellite Launch Providers... 6
Chapter 2: Inside Virgin Orbit – A New Space Venture..11
Chapter 3: The Launch System – Cosmic Girl and LauncherOne.. 16
Chapter 4: The Air Launch Advantage...................... 23
Chapter 5: LauncherOne – The Simplistic and Scalable Rocket.. 27
Chapter 6: A Milestone for Virgin Orbit – First Successful Launches...32
Chapter 7: Expanding the Mission – Plans for Growth.. 38
Chapter 8: The Broader Context – The New Space Race for Small Satellite Launches............................ 44
Chapter 9: Applications and Future Potential of Small Satellites... 51

Introduction

Virgin Orbit emerged as a unique player in the expanding space industry, standing out with a distinctive approach to getting satellites into orbit. Instead of the traditional ground launches most are familiar with, Virgin Orbit has redefined the launch process by using an aircraft to carry its rockets partway to space. This innovative method—known as air launch—brings a fresh perspective to satellite deployment, blending aviation with rocketry in a way that few imagined.

This venture is often mistaken for its sister company, Virgin Galactic, which has captured headlines with space tourism flights that let civilians experience the thrill of space travel. While Virgin Galactic is designed for those who dream of reaching the edge of space for a few minutes, Virgin Orbit is all about business and utility. The company's primary mission revolves around deploying small satellites quickly, flexibly, and affordably. It aims to meet the demand of governments, commercial entities, and scientific researchers who need their technology in orbit without the long wait times or compromises associated with traditional launch methods.

Despite operating under the same Virgin brand, Virgin Orbit only recently began to carve out its own public identity. When it went public on the New York Stock Exchange, the company made its mark by parking a rocket right in the middle of Times Square, signaling its ambition and intent. This bold gesture, alongside its groundbreaking launch method, has propelled Virgin Orbit into the spotlight. With its planes as mobile launchpads, it's introducing the world to a new era of space access, blending the efficiency of air travel with the power of rocketry to meet a growing need in an increasingly connected world.

Chapter 1: The Rise of Small Satellite Launch Providers

In recent years, a quiet yet transformative shift has unfolded in the world of satellite technology. Where satellites once took up the size of small buses, today's advances in miniaturization have reduced these machines to dimensions as compact as a shoebox or even smaller. These advancements mark the beginning of a new "small satellite revolution," driven by the desire for nimble, efficient, and less costly devices that can still achieve sophisticated tasks in orbit. Small satellites have grown increasingly capable, offering communication, imaging, environmental monitoring, and research functions that were once exclusive to much larger and costlier counterparts.

The demand for deploying these compact satellites is skyrocketing, prompting a need for launch services that can keep pace with the frequency and affordability that customers expect. While traditional rockets were built to carry massive payloads, they rarely meet the exact needs of small satellite operators, who often end up as secondary payloads on larger missions. For companies needing to deploy their technology swiftly and on precise schedules, ride-sharing on larger rockets often isn't a viable option. This is where dedicated small-satellite launch providers have stepped in, recognizing a significant gap in the market.

The rise of companies like Rocket Lab, Firefly Aerospace, and Astra has fueled a new wave in the space industry, each company leveraging unique methods and technology to meet the demand for affordable, small-payload launches. Rocket Lab, for instance, uses its Electron rocket, a vehicle specifically designed for lightweight payloads. It has already achieved a notable foothold in the market by offering reliable, frequent launches and was among the earliest to tap into this niche. Firefly, meanwhile, is developing rockets aimed at carrying slightly larger payloads, hoping to capture both small and medium satellite clients. Astra, known for its rapid, streamlined manufacturing process, focuses on creating rockets that are simple, cost-effective, and ideal for frequent launches.

Together, these companies reflect the larger shift in space technology, where smaller, more agile players are emerging to meet the specialized needs of an evolving market. As this revolution continues, the landscape of satellite deployment is undergoing a seismic transformation, one that is democratizing access to space and bringing powerful new tools to organizations across the globe.

Virgin Orbit occupies a distinct position within the rapidly evolving small satellite launch industry, setting itself apart through a groundbreaking approach that combines the flexibility of aviation with the power of rocketry. While many new space companies are developing lightweight rockets to launch from the ground, Virgin Orbit takes a different path by releasing its rockets mid-air from a modified Boeing 747. This air-launch technique allows Virgin Orbit to sidestep the complexities and constraints of traditional ground-based launches, offering a flexible, cost-effective alternative for getting payloads into orbit.

The market demand for small satellite launches has grown exponentially, spurred by sectors that rely on satellites for everything from telecommunications and environmental monitoring to national security and research. Companies and governments increasingly need quick, reliable access to space to keep pace with technological advancements and stay competitive. Virgin Orbit addresses this demand by offering a "launch on-demand" capability. Unlike fixed launch sites, which are often booked months or even years in advance, Virgin Orbit's mobile platform allows them to deploy from any major runway around the world. This flexibility means that they can launch on relatively short notice, delivering satellites to specific orbits that might be challenging to reach from traditional launch pads.

For clients with time-sensitive projects or those seeking more control over their satellite's destination in orbit, Virgin Orbit's approach represents a welcome shift. By using a plane as a launch platform, Virgin Orbit opens up new possibilities for orbital inclinations and launch sites, allowing for a tailored service that many satellite operators find invaluable. This method not only reduces the logistical barriers of scheduling and location but also minimizes fuel consumption and costs by taking advantage of the altitude and speed of the 747 before releasing the rocket. Virgin Orbit is not just another small satellite launch provider; it's pioneering a new category of launch service designed to meet the needs of a dynamic and fast-growing industry.

Chapter 2: Inside Virgin Orbit – A New Space Venture

Virgin Orbit traces its roots back to the larger Virgin Group, the iconic brand founded by Sir Richard Branson. Known for ventures that range from airlines to media companies, Virgin has always had a reputation for pushing boundaries and challenging norms, and its expansion into space is no different. Virgin Orbit originated as an offshoot of Virgin Galactic, the space tourism company that Branson established with the ambition of bringing civilians to the edge of space. However, as the needs of the aerospace market began to diversify, it became clear that there was an opportunity for a separate venture dedicated to launching small satellites.

Branson's vision for Virgin Orbit was inspired by the growing importance of small satellites in industries such as communications, environmental monitoring, and defense. Recognizing the constraints of traditional launch services, which often require long lead times and limit flexibility, he saw an opportunity to make satellite deployment faster and more accessible. Branson has always been a vocal proponent of innovation, and he believed that Virgin Orbit could introduce a game-changing approach by integrating aviation with space launch technology. This led to the development of the air-launch system that would eventually become Virgin Orbit's defining feature.

From the outset, Branson's approach to Virgin Orbit embodied the Virgin ethos of bold, forward-thinking projects designed to disrupt established markets. Under his leadership, Virgin Orbit developed LauncherOne, a custom-designed rocket that could be deployed from Cosmic Girl, a repurposed Boeing 747. This unconventional method exemplified Branson's commitment to rethinking how space access could work in the modern world. By leveraging Virgin's expertise in both aviation and innovation, Branson envisioned Virgin Orbit as a solution for industries and governments needing efficient, flexible, and on-demand access to space. His goal for Virgin Orbit goes beyond mere business; he envisions it as part of a broader effort to democratize space access, enabling a range of sectors to harness the power of satellites without the prohibitive costs or delays of traditional launch methods.

Virgin Orbit is thus not just another branch of the Virgin empire but a project that aligns closely with Branson's lifelong mission of breaking down barriers and expanding human potential. Under his vision, Virgin Orbit aims to make space a realm that's accessible, practical, and transformative for industries around the world.

Virgin Orbit's primary mission revolves around offering flexible and dedicated satellite launch services tailored to the unique needs of both commercial and government clients. In a rapidly advancing world where satellite technology plays an essential role in everything from communication and data gathering to defense and environmental monitoring, Virgin Orbit steps in to provide a timely, adaptable, and affordable path to orbit. The company's approach is focused on making space more accessible by eliminating many of the logistical constraints associated with traditional launch services.

For commercial clients, Virgin Orbit delivers a streamlined process that enables quick deployment, customized scheduling, and precise orbit placement—crucial factors for industries relying on data networks, global positioning, and imaging technology. By using its mobile, air-launched rocket system, Virgin Orbit offers a way to launch from any suitable runway, which means clients no longer have to depend on crowded or geographically limited spaceports. This flexibility not only reduces wait times but also allows for launches to specific orbits that may otherwise be challenging to reach.

Government clients also benefit significantly from Virgin Orbit's capabilities. National security, weather monitoring, and global communication often require dedicated, reliable, and controlled access to space. Virgin Orbit's ability to launch on short notice and from various global locations provides a level of responsiveness that is increasingly valuable for government agencies. By offering a dependable, dedicated service for small satellite deployment, Virgin Orbit allows these organizations to meet critical mission needs without the constraints of traditional rocket scheduling and capacity.

At its core, Virgin Orbit's mission is to democratize access to space by addressing the needs of today's fast-paced, technology-driven world. By bridging aviation and aerospace with a flexible and dedicated launch service, Virgin Orbit is paving the way for a new era in space access—one that aligns with the growing demands of industries and government agencies worldwide.

Chapter 3: The Launch System – Cosmic Girl and LauncherOne

Virgin Orbit's launch system is defined by two key components: Cosmic Girl, a modified Boeing 747, and LauncherOne, a purpose-built two-stage rocket. Together, they form an unconventional yet highly effective platform for delivering small satellites to space, blending the versatility of a jet with the power of rocketry in a way that few others have attempted.

Cosmic Girl, originally a commercial passenger aircraft, has been repurposed and customized to serve as the launch platform for Virgin Orbit's missions. This 747 jet is engineered to carry LauncherOne beneath its left wing, much like a missile attached to a fighter jet. By releasing the rocket mid-air at an altitude of approximately 35,000 feet, Cosmic Girl effectively gives LauncherOne a significant head start, allowing the rocket to conserve fuel and carry more payload to its destination. The 747's expansive range and ability to take off from virtually any major runway mean that Virgin Orbit can operate launches from various global locations, offering unmatched flexibility in terms of launch sites and orbital inclinations.

The second essential component, LauncherOne, is a two-stage orbital rocket engineered specifically for small satellite payloads. Designed for simplicity and efficiency, this rocket has only one engine in each stage, a streamlined structure that contrasts sharply with the complex multi-engine configurations of traditional rockets. LauncherOne's first stage propels it into the upper atmosphere, where the second stage takes over, boosting the payload into its intended orbit. Capable of carrying up to 500 kilograms, this two-stage system is optimized for efficiency, making it an ideal solution for the demands of modern small satellite missions.

Together, Cosmic Girl and LauncherOne redefine the launch process by creating a mobile, adaptable, and cost-effective system that can reach space from any viable runway, providing an alternative to traditional ground-based launches. This approach allows Virgin Orbit to serve a diverse range of clients, offering dedicated, on-demand access to space through a seamless blend of aviation and rocketry.

The transformation of Cosmic Girl from a commercial Boeing 747 into a critical part of Virgin Orbit's launch system is a feat of engineering designed to marry aviation with rocketry. To adapt this passenger aircraft for a unique role in launching rockets, Virgin Orbit undertook extensive modifications aimed at optimizing both weight and structural capacity. The interior of Cosmic Girl was stripped of non-essential components to reduce weight, creating a leaner aircraft capable of carrying LauncherOne and the additional mission control systems needed for high-altitude rocket deployment.

One of the most critical adjustments was the installation of a specialized pylon on the underside of Cosmic Girl's left wing, just outside the inboard engine. This pylon serves as the attachment point for LauncherOne, holding the rocket securely during ascent while being engineered to handle the additional load of a multi-ton payload. This setup is reminiscent of fighter jet technology, where missiles or other heavy equipment are mounted under the wings. The pylon and attachment system are designed to release the rocket cleanly, allowing it to free-fall for a brief moment before its engine ignites, launching it further into space.

Moreover, Cosmic Girl includes a mobile mission control system onboard. This enables Virgin Orbit's team to monitor all aspects of the mission from within the aircraft, coordinating every detail of the launch sequence and ensuring precise timing for LauncherOne's release. The combination of these modifications enables Cosmic Girl to act as a high-altitude launch platform, giving LauncherOne the initial lift it needs to reach space while conserving fuel that would otherwise be used in a ground-based launch.

This integration of aviation and rocketry showcases the ingenuity behind Virgin Orbit's approach, adapting existing technology in new ways to achieve efficiency and flexibility. By modifying Cosmic Girl to support LauncherOne, Virgin Orbit has created a launch system that operates with the precision of aerospace engineering while offering the adaptability of an aircraft—a powerful blend that has the potential to reshape how satellites are delivered to space.

Cosmic Girl is not only a launch platform but also a mobile command center, equipped with a sophisticated onboard mission control system that enables real-time monitoring and management of each launch. This system allows Virgin Orbit's engineers and mission specialists to operate directly from within the aircraft, managing the intricacies of the launch as it unfolds thousands of feet above the Earth. By incorporating mission control directly into Cosmic Girl, Virgin Orbit can streamline operations and respond to any last-minute variables with precision, making the process both efficient and adaptive.

The mobile mission control setup includes a suite of computers, communication systems, and telemetry instruments, all configured to keep tabs on LauncherOne's status. From the moment Cosmic Girl takes off, the mission control team monitors every aspect of the launch vehicle, from fuel levels and temperatures to structural integrity and positioning. This in-flight setup provides real-time feedback on LauncherOne's readiness, tracking its performance under the wing and preparing for the precise moment of release.

One of the critical functions of this system is to manage the countdown sequence and ensure that all parameters are optimal for launch. The mission control team synchronizes the release of LauncherOne with factors such as altitude, speed, and the aircraft's orientation, as well as external conditions like weather and wind speeds. Once LauncherOne is released, the system continues to monitor telemetry from the rocket, providing data on its trajectory and engine performance as it ascends into orbit.

This onboard mission control capability gives Virgin Orbit an operational edge, allowing the team to conduct the entire launch sequence from start to finish without the need for extensive ground support. By embedding mission control within the aircraft, Virgin Orbit achieves a level of mobility and responsiveness that's uncommon in the launch industry, enabling the company to operate from different locations worldwide while maintaining strict control over each stage of the mission.

Chapter 4: The Air Launch Advantage

Launching rockets from an aircraft, as Virgin Orbit does, introduces several strategic advantages over traditional ground-based launches, fundamentally reshaping how satellites are delivered to space. This approach enhances fuel efficiency, expands operational flexibility, and enables faster, more targeted launches, each of which addresses key challenges in modern satellite deployment.

One major advantage is **fuel efficiency**. By carrying the rocket to an altitude of 35,000 feet, Virgin Orbit's Cosmic Girl 747 bypasses a significant portion of the dense lower atmosphere that ground-launched rockets must push through. This altitude jump means that LauncherOne requires less fuel to reach the upper atmosphere, where the rocket's journey into orbit truly begins. This fuel savings translates to an increase in payload capacity, allowing Virgin Orbit to carry more cargo weight relative to the rocket's size. The reduced fuel requirement not only makes each launch more efficient but also enables the rocket to accommodate heavier payloads, offering greater value to clients within the constraints of a smaller, cost-effective vehicle.

Flexible launch locations represent another transformative advantage. Unlike traditional rocket launches, which are restricted to fixed launch pads with specific geographic constraints, Virgin Orbit's air-launch method allows them to take off from any major airport with a suitable runway. This flexibility means that Virgin Orbit can offer launches from a variety of locations worldwide, enabling access to a broader range of orbits and inclinations. For example, launching closer to the equator or adjusting the takeoff location can optimize the rocket's trajectory and minimize flight distance, allowing for precise targeting of specific orbital paths, including those challenging to reach from ground-based sites. This freedom to launch from diverse locations not only expands Virgin Orbit's operational range but also supports unique mission requirements, positioning the company as an adaptable and responsive launch provider.

In addition, Virgin Orbit's system allows for **timely and targeted launches**. Traditional launches often require satellite operators to wait for available slots on larger rockets, sharing space with other payloads and relinquishing control over timing and destination. Virgin Orbit, by contrast, offers dedicated, custom launch windows that allow clients to deploy their technology on short notice. This capability is especially valuable for companies and government agencies with time-sensitive missions, where delays could impact research, communication networks, or security measures. By providing a tailored launch option, Virgin Orbit lets clients launch according to their own schedules, deploying payloads at the precise moment needed to optimize satellite function.

These combined advantages—fuel efficiency, global flexibility, and rapid, targeted deployment—highlight the ingenuity behind Virgin

Orbit's air-launch approach. By blending the mobility of aviation with the power of rocketry, Virgin Orbit opens new possibilities for satellite operators, meeting modern demands with an agile, efficient, and forward-thinking system that redefines access to space.

Chapter 5: LauncherOne – The Simplistic and Scalable Rocket

LauncherOne, Virgin Orbit's dedicated small satellite rocket, is designed with a two-stage, disposable structure optimized for efficiency, simplicity, and cost-effectiveness. In contrast to the trend of developing reusable rockets, Virgin Orbit chose to make LauncherOne a fully expendable vehicle. This decision reflects the company's focus on keeping launch costs low and maintaining a streamlined design that allows for rapid manufacturing and scalability. By making the rocket disposable, Virgin Orbit reduces the engineering and maintenance demands that come with reusable systems, ensuring that each LauncherOne mission is a fresh start with fewer complications and a shorter turnaround time.

LauncherOne's design stands out for its simplicity, featuring only one engine in each of its two stages. This minimalist approach contrasts with many other small satellite rockets, which often use multi-engine configurations to achieve higher lift capabilities. For example, Rocket Lab's Electron rocket has nine engines on its first stage alone, creating a complex system that requires extensive coordination between components. By comparison, LauncherOne's two-engine design reduces the risk of mechanical failure and minimizes the need for intricate control systems. Fewer components also mean a lighter overall structure, allowing more of the rocket's capacity to be dedicated to payload weight rather than structural bulk.

This simplicity extends to LauncherOne's production as well. Virgin Orbit builds the rocket entirely in-house at its California facility, where the streamlined design allows for rapid construction and assembly. This "factory-to-launch" approach enables Virgin Orbit to maintain control over quality and production timelines, providing clients with a reliable and efficient option for getting their technology into orbit.

Overall, LauncherOne's two-stage, disposable structure exemplifies Virgin Orbit's commitment to straightforward, cost-effective satellite deployment. With fewer components and a simpler design than many competitors, it delivers a pragmatic approach that aligns with the specific needs of the small satellite market—offering a focused solution for clients who value affordability, reliability, and timely access to space.

Virgin Orbit's LauncherOne rockets are manufactured from the ground up at the company's dedicated facility in Long Beach, California. This centralized plant, where every step of production—from initial assembly to final testing—occurs under one roof, embodies Virgin Orbit's commitment to efficiency, control, and rapid scalability. By streamlining the entire manufacturing process within a single location, Virgin Orbit can closely manage quality, refine production techniques, and optimize timelines, all essential for meeting the increasing demand for small satellite launches.

The California plant's approach brings significant advantages in terms of **production speed**. With every stage of the rocket's creation taking place in one location, Virgin Orbit reduces logistical complexities that might otherwise arise from transporting parts between different sites or relying on third-party vendors for critical components. This streamlined setup allows engineers to quickly address any issues, make adjustments as needed, and move each LauncherOne unit seamlessly from assembly to testing. The result is a faster production cycle, enabling Virgin Orbit to meet tight launch schedules and respond promptly to customer demand.

This facility also positions Virgin Orbit for **scalability**. As demand for small satellite launches grows, Virgin Orbit can scale up production without overhauling its processes or expanding to multiple sites. The flexibility of the in-house plant allows the company to ramp up production volumes in a controlled manner, preparing for its goal of producing and launching up to 20 LauncherOne rockets per year. This centralized approach gives Virgin Orbit a unique edge in the market, allowing it to offer clients both reliability and the ability to adapt quickly to increased launch needs without compromising on quality.

By manufacturing each LauncherOne under one roof, Virgin Orbit benefits from a level of agility that's uncommon in the aerospace industry. This setup enables the company to deliver a steady, predictable flow of rockets ready for launch, making it a dependable choice for clients who need a reliable path to orbit and reinforcing Virgin Orbit's reputation as an agile and responsive player in the modern space landscape.

Chapter 6: A Milestone for Virgin Orbit – First Successful Launches

Virgin Orbit's first successful mission in January 2021 marked a significant milestone for the company, proving that its air-launch system could deliver payloads to orbit. On this mission, LauncherOne carried ten CubeSats into low Earth orbit, demonstrating its capability to handle the demands of modern small satellite deployment. These CubeSats—miniature satellites designed for a variety of applications in research, communication, and data collection—represented the type of payloads that LauncherOne was built to carry. This mission validated Virgin Orbit's approach, showcasing the effectiveness of launching rockets from high altitudes and providing a glimpse into the company's potential to meet the needs of clients worldwide.

Following this success, Virgin Orbit completed its first operational mission, titled "Tubular Bells Part 1," in June 2021. This mission carried a diverse payload of seven satellites for clients including the U.S. Department of Defense, Polish satellite company SatRevolution, and the Royal Netherlands Air Force, which launched its first military satellite. While this mission was commercially significant, its name held a sentimental nod to Virgin's history. "Tubular Bells" was the first album released by Virgin Records, the company Richard Branson founded in 1973, launching his entrepreneurial career. The album, an instrumental piece by musician Mike Oldfield, became an unexpected hit and cemented Virgin Records as a key player in the music industry.

By naming the mission "Tubular Bells Part 1," Virgin Orbit honored Branson's early success and the legacy of innovation that has been central to the Virgin brand. This mission not only advanced Virgin Orbit's credibility in the aerospace sector but also tied its present achievements to the entrepreneurial spirit that launched Virgin itself. "Tubular Bells Part 1" became more than just a launch—it was a symbolic fusion of Virgin's past with its vision for the future, marking a new chapter in Branson's journey from music mogul to space innovator. Through these missions, Virgin Orbit has proven itself as a reliable and innovative force, dedicated to expanding space access and continuing the Virgin legacy of daring to disrupt the norm.

The "Tubular Bells Part 1" mission in June 2021 carried a diverse and strategically significant array of payloads, reinforcing Virgin Orbit's operational capabilities and solidifying its credibility in the space industry. Among the seven satellites on board, the payload included research, commercial, and military satellites, each designed for specific missions in low Earth orbit, ranging from defense to imaging and scientific research. This mission represented Virgin Orbit's commitment to offering flexible, dedicated services that meet a variety of client needs.

One of the primary payloads on this mission came from the U.S. Department of Defense, which included four research and development CubeSats. These satellites were launched for various experimental purposes, demonstrating Virgin Orbit's ability to support defense-related missions that require precise deployment and reliable performance. For the Department of Defense, using Virgin Orbit was an opportunity to test new technologies and gain critical data on the effectiveness of the air-launch system for national security applications. Successfully completing this deployment strengthened Virgin Orbit's reputation as a dependable partner for government agencies and set a precedent for future defense collaborations.

The mission also carried two optical imaging satellites for SatRevolution, a Polish satellite technology company specializing in Earth observation. These satellites are part of SatRevolution's ongoing project to establish a constellation that provides high-resolution images for applications like environmental monitoring, urban planning, and resource management. By delivering these satellites to their intended orbit, Virgin Orbit demonstrated its capability to serve commercial clients in industries that rely on timely, detailed Earth data. The success of these deployments highlighted LauncherOne's suitability for companies in the growing satellite imaging market, showcasing the rocket's precision and dependability.

Another key payload was a milestone for the Royal Netherlands Air Force, which launched its first-ever military satellite on this mission. This satellite represents a step forward in the Netherlands' commitment to utilizing space for national defense, with applications that could include secure communications and reconnaissance. Successfully carrying this satellite into orbit validated Virgin Orbit's ability to meet military specifications and requirements, adding to its appeal as a launch provider for nations looking to expand their space capabilities.

The range and diversity of the payloads on the "Tubular Bells Part 1" mission underscored Virgin Orbit's versatility and reliability in handling different types of satellite technology. Each successful deployment served as an operational proof-of-concept, showing clients across sectors that Virgin Orbit's air-launch system is not only viable but highly capable. By demonstrating its capacity to support both commercial and defense missions, Virgin Orbit established itself as a credible force in the small satellite launch market, prepared to meet the growing demand for dedicated, responsive launch solutions.

Chapter 7: Expanding the Mission – Plans for Growth

Virgin Orbit has set ambitious goals for growth, aiming to scale its launch operations to 20 missions annually, with the long-term vision of further increasing this frequency to meet rising demand. This target reflects the company's commitment to becoming a premier launch provider for small satellites, particularly in a market where clients seek flexibility, reliability, and speed to orbit. To achieve this level of operational scalability, Virgin Orbit is focusing on optimizing its production, streamlining launch processes, and expanding partnerships with both commercial and government sectors.

A significant component of this plan involves the ongoing development of Virgin Orbit's manufacturing and operational infrastructure. By maintaining an efficient, in-house production line for LauncherOne at its Long Beach facility, Virgin Orbit ensures a steady and scalable supply of rockets ready for launch. This approach allows the company to quickly respond to client demands, particularly as it anticipates an increase in missions that require rapid deployment or custom launch windows. Virgin Orbit's mobile launch system, enabled by its air-launch capability, is another asset that supports scalability; by taking off from various locations globally, it can quickly adapt to new market needs without relying on fixed launch sites.

Virgin Orbit's goal to reach 20 launches per year also aligns with the growing applications of small satellite technology, from telecommunications and Earth observation to national security. By scaling up its operations, Virgin Orbit positions itself to support a broader range of clients across industries, providing dedicated launches that cater to specific orbital paths and schedules. The company's focus on increasing launch frequency reflects its dedication to being a responsive, go-to partner in an industry where timely access to space is critical.

In the long term, Virgin Orbit aims not only to meet this 20-launch target but to exceed it, contributing to the larger ecosystem of space services and helping advance the possibilities of satellite technology. Through continuous improvements in production, operational agility, and launch availability, Virgin Orbit is steadily building the foundation to become a central player in the expanding landscape of small satellite launches.

The market potential for small satellite launches is on a rapid ascent, driven by the growing need for real-time data, global connectivity, and enhanced security capabilities across various industries. As technology advances, small satellites are becoming powerful tools for tasks that once required much larger, more expensive payloads. Industries like telecommunications, Earth observation, climate monitoring, and defense now rely heavily on these compact satellites to deliver critical data, improve communications, and enhance surveillance. This shift has created a substantial demand for cost-effective, dedicated launch services that can adapt to the unique needs of these industries.

In the coming years, demand for small satellite launches is expected to surge as more organizations across commercial and governmental sectors adopt satellite technology. Telecommunications companies, for instance, are investing in satellite constellations to bring internet access to remote areas, bridging connectivity gaps worldwide. Similarly, Earth observation and environmental monitoring are becoming essential for tracking climate change, managing natural resources, and responding to natural disasters. Governments and defense agencies also rely increasingly on small satellites for secure communications and reconnaissance, supporting national security and infrastructure resilience. This widespread interest is projected to fuel a vibrant market for small satellite launches, with frequent missions becoming the norm as organizations seek regular access to space.

Virgin Orbit's air-launch system places it in a unique position to capitalize on this growing demand. Unlike ground-based launch providers that operate from fixed spaceports, Virgin Orbit can take off from a variety of global locations, giving it an operational flexibility that traditional launches cannot match. This capability is especially attractive for clients in both commercial and governmental sectors who need rapid deployment to specific orbits on short notice. For telecommunications providers or environmental agencies, the ability to launch quickly and from strategic locations enhances Virgin Orbit's appeal as a reliable partner for meeting real-time demands.

The air-launch method also enables Virgin Orbit to avoid many of the scheduling conflicts and delays associated with shared launch pads, giving clients greater control over their missions. In the governmental sector, where timing and security are paramount, Virgin Orbit's flexibility and ability to launch at any time from different locations provide an additional layer of strategic advantage. This adaptability positions Virgin Orbit as a responsive and accessible solution for clients across sectors, uniquely equipping the company to support the surge in small satellite deployment that is reshaping the landscape of space technology.

With a system designed for rapid, tailored access to space, Virgin Orbit is prepared to be a primary driver in the new era of small satellite launches, offering a distinct service that meets the evolving needs of industries seeking to leverage space in impactful ways.

Chapter 8: The Broader Context – The New Space Race for Small Satellite Launches

The small satellite market has given rise to a "new space race," where a dynamic mix of companies are competing to become the preferred choice for launching compact, cost-effective payloads into orbit. This competition reflects a shift from the traditional model of space exploration and high-stakes missions to a market-driven approach that prioritizes accessibility, flexibility, and rapid deployment. Companies like Virgin Orbit, Rocket Lab, Astra, and Firefly Aerospace have entered this race, each developing unique solutions to meet the soaring demand for small satellite launches. Their focus isn't on massive rockets or deep-space exploration but on practical, efficient systems that can frequently deliver satellites to low Earth orbit. This modern space race is fueled by a need to serve industries that rely on fast data, global connectivity, and secure communication—all of which are becoming essential elements of the digital economy.

This emerging competition stands in stark contrast to the high-profile rivalry between SpaceX and Blue Origin. While both companies are notable players in space, their ambitions are set on the larger scale of interplanetary missions and deep-space exploration, as well as securing lucrative NASA contracts for missions to the Moon and beyond. The competition between SpaceX and Blue Origin has centered around securing top-tier contracts with NASA for flagship missions, including the Artemis program, which aims to return humans to the lunar surface. These endeavors require large, reusable rockets, complex spacecraft, and extensive infrastructure, all designed for exploration and potential human settlement beyond Earth.

In the small satellite launch market, however, the goal is agility rather than scale. Companies like Virgin Orbit focus on providing tailored, on-demand access to space for clients who need regular, dedicated launches for their technology. The emphasis is on creating flexible launch options that can accommodate frequent, small-scale missions, allowing for a steady stream of satellites to enter orbits that support telecommunications, Earth monitoring, and scientific research. This approach doesn't require the heavy-lift rockets or deep-space capabilities that SpaceX and Blue Origin are building; instead, it relies on innovations like air-launch technology and streamlined, disposable rockets that are optimized for cost and speed.

Thus, while SpaceX and Blue Origin battle over contracts for ambitious, long-term exploration projects, the small satellite market thrives on responsive, specialized services that open space to a wider range of industries and applications. This new space race is more about utility and efficiency than exploration, representing a pragmatic shift in the commercial space industry that is transforming how we access and use space.

Small satellite providers are pivotal in driving the next wave of technological advancements in space, opening up a world of possibilities that extend far beyond traditional applications. By providing affordable, frequent, and flexible access to space, these providers enable a broad range of industries to integrate satellite technology into their operations, fueling innovation across fields like telecommunications, environmental science, national security, and beyond. Their role goes beyond simply putting satellites into orbit; they are actively shaping a new space economy that leverages satellites for real-time data, global connectivity, and infrastructure resilience.

One of the most transformative impacts of small satellite providers is in the realm of **telecommunications and global internet access**. Companies like SpaceX's Starlink and other emerging satellite constellations rely on small satellites to create networks that can deliver high-speed internet to remote and underserved regions. These constellations, launched by small satellite providers, promise to bridge the digital divide, offering internet access in places where terrestrial infrastructure is either limited or nonexistent. This access can boost education, healthcare, and economic opportunities for millions of people, accelerating global connectivity in ways previously unimaginable.

In **Earth observation and environmental monitoring**, small satellites provide invaluable data on climate patterns, natural resource management, and disaster response. Satellites equipped with imaging and sensor technology can track deforestation, monitor water resources, and detect changes in polar ice, contributing to a deeper understanding of global environmental health. These data-driven insights support proactive measures in climate adaptation and resource conservation, as well as rapid response during natural disasters, where real-time satellite data can guide relief efforts.

National security and defense are also evolving with the help of small satellite technology.

Government agencies are leveraging these compact, rapidly deployable satellites for secure communications, surveillance, and reconnaissance. Unlike large, traditional satellites that may take years to develop and deploy, small satellites can be launched quickly and replaced frequently, enabling defense and intelligence agencies to respond to emerging threats and adapt their surveillance capabilities on short notice.

Small satellite providers further advance **technological experimentation and innovation** by enabling universities, research institutions, and startups to send their projects to space. These smaller entities, which might lack the budgets for traditional launches, can now deploy experimental satellites and test groundbreaking technologies in the low-gravity environment of space. This democratized access to orbit is fostering a new generation of research and development that has applications in fields as diverse as biomedicine, materials science, and robotics.

By expanding the possibilities for space access, small satellite providers are playing a critical role in building an infrastructure that supports emerging technologies and global needs. They are the engine behind a wave of innovation that brings space closer to everyday life, allowing organizations across the world to harness space in ways that benefit society, protect the planet, and drive technological progress forward.

Chapter 9: Applications and Future Potential of Small Satellites

Small satellite technology is at the forefront of a shift in space-based applications, enabling breakthroughs across industries that rely on data, connectivity, and real-time insights. From expanding global internet access to enhancing national security, these compact satellites have become versatile tools that make space accessible and practical for addressing global needs.

In **low-latency internet and telecommunications**, small satellite constellations are transforming the potential for global connectivity. By positioning thousands of satellites in low Earth orbit, companies can establish networks that provide fast, reliable internet to even the most remote areas. These low-latency networks bypass the geographical limitations of traditional infrastructure, delivering high-speed connectivity to communities with limited or no access to conventional broadband. For remote villages, offshore industries, and disaster-prone areas, small satellite networks can become lifelines, supporting healthcare, education, business, and emergency communications. This expanding connectivity opens up new economic and social opportunities, bridging the digital divide and contributing to a more connected world.

In **climate and environmental monitoring**, small satellites play a vital role in tracking and analyzing environmental changes. Equipped with advanced sensors and imaging systems, these satellites can monitor deforestation, ice melt, air quality, and ocean temperatures, providing detailed data that enhances our understanding of climate change and helps organizations make data-driven decisions. Small satellites contribute valuable insights for managing natural resources, predicting weather patterns, and responding to natural disasters. For instance, real-time data on wildfires, floods, and hurricanes allows emergency services to prepare and respond more effectively, reducing loss of life and property damage. In agriculture, environmental data from small satellites enables farmers to track soil health, crop conditions, and water availability, optimizing yield and promoting sustainable practices.

National security and defense applications have also expanded with the use of small satellites, offering governments new capabilities in surveillance, secure communications, and intelligence gathering. Unlike traditional, large-scale defense satellites that take years to design and launch, small satellites can be developed and deployed quickly, providing governments with more flexible and cost-effective surveillance options. These satellites are ideal for monitoring border activity, tracking the movement of fleets, and observing strategic areas of interest in real-time. With their ability to quickly refresh and replace satellite constellations, defense agencies can respond to emerging threats and adapt their strategies on short notice, maintaining a competitive advantage in intelligence and security operations.

In **supply chain and asset tracking**, small satellites have introduced transformative solutions for logistics and real-time monitoring. By connecting to IoT devices, small satellites offer businesses the ability to track shipments, monitor assets, and manage fleets worldwide, even in remote or maritime locations that lack traditional connectivity. This capability enables supply chains to operate with greater transparency and resilience, allowing companies to monitor cargo conditions, manage inventory, and respond swiftly to disruptions. For industries like agriculture, energy, and shipping, where assets often travel through remote or harsh environments, satellite tracking provides real-time data that supports efficient logistics and proactive risk management.

The anticipated growth of the small satellite industry signals a broader evolution in the space sector as it adapts to meet these emerging demands. By delivering solutions that serve telecommunications, environmental science, defense, and logistics, small satellite providers are reshaping our understanding of space's practical uses. This growth is set to accelerate as new innovations, such as miniaturization and enhanced data processing capabilities, push the limits of what small satellites can achieve. As the space industry evolves, small satellites will likely play an ever-increasing role in addressing global challenges, making space technology more integral to our everyday lives and driving economic development in the years to come. The small satellite revolution thus marks the beginning of a more accessible, versatile, and data-driven era in space, one that holds promise for innovation across nearly every industry.

Conclusion

Virgin Orbit's journey from concept to execution represents a bold reimagining of how satellites reach space. What began as an innovative idea within the Virgin Group has evolved into a pioneering launch system that combines aviation and rocketry in ways few could have anticipated. With its air-launch approach, Virgin Orbit has carved out a unique space in the industry, providing clients with a level of flexibility, speed, and affordability that challenges traditional launch methods. By transforming a Boeing 747 into a mobile launchpad and perfecting the design of the LauncherOne rocket, Virgin Orbit has harnessed a streamlined solution that caters to the burgeoning demand for small satellite deployments.

The long-term impact of air-launched rockets could be transformative, especially for smaller clients who need dedicated, custom launch services without the prohibitive costs or scheduling constraints of larger rockets. Virgin Orbit's air-launch method not only opens up more geographic options for launches but also provides on-demand access to space, empowering businesses, research institutions, and governments alike to deploy their technology when and where they need it. For smaller satellite operators and sectors with time-sensitive missions, the ability to launch quickly and flexibly makes a critical difference, enabling them to stay competitive in a data-driven world.

Looking ahead, Virgin Orbit and companies following similar models are well-positioned to reshape the space economy. By democratizing access to space through innovative approaches, they are lowering the barriers for organizations that once faced prohibitive costs and long wait times to bring their ideas to orbit. As this sector continues to grow, the impact of these companies will likely extend far beyond space technology, influencing connectivity, environmental sustainability, national security, and scientific advancement on Earth. Virgin Orbit stands at the forefront of a new era in space, one that brings the power of orbit within reach for a wider range of industries, unlocking possibilities that could redefine our relationship with space and its role in our daily lives.

www.ingramcontent.com/pod-product-compliance
Lightning Source LLC
Chambersburg PA
CBHW050311220526
45465CB00005B/1944